This "Duffy" Book
was generously provided by
Corporate Partners:

CaroTrans
and
MAINFREIGHT USA

www.DuffyBooksInHomesUSA.org

PowerPhonics™

Clouds

Learning the CL Sound

Susan Tanner

The Rosen Publishing Group's
PowerKids Press™
New York

Look at the sky. Do you see clouds?

Clouds can look clean and white.

Clouds can look gray.

Clouds can be way up in the sky.

Clouds can be close by.

Clouds can make snow.

Clouds can make rain.

Clouds can clear away.

There are no clouds.

The sky is clear. Good-bye, clouds!

Word List

clean

clear

close

clouds

Instructional Guide

Note to Instructors:
One of the essential skills that enable a young child to read is the ability to associate letter-sound symbols and blend these sounds to form words. Phonics instruction can teach children a system that will help them decode unfamiliar words and, in turn, enhance their word-recognition skills. We offer a phonics-based series of books that are easy to read and understand. Each book pairs words and pictures that reinforce specific phonetic sounds in a logical sequence. Topics are based on curriculum goals appropriate for early readers in the areas of science, social studies, and health.

Letter/Sound: cl – On a chalkboard or dry-erase board, list familiar words beginning with consonant l: *lap, lamp, lay, lean, lump, lock, lip, luck*. Have the child add **c** to the beginning of each word to create a new word with the initial consonant blend **cl**. Help them decode the new words.
- Ask the child to find the new word that rhymes with *lap*. Continue with the remaining words. Have the child use the new words in sentences.

Phonics Activities: Introduce members of the Clown Club (clown faces made from paper plates), whose members have names beginning with **cl**. Give two names for each clown. Have the child select the one that begins with **cl**. (Examples: *Buttons* or *Cleo*, *Clancy* or *Dooley*.) Write the clown names. Have the child underline the **cl** blends and choose one of the clown faces. Pronounce several words, some with the initial **cl** blend (vocabulary words and words from activity 1) and some with **hard c** or **l** beginning sounds. Have the child show their clown face when they hear a **cl** word.
- Buy or make flash cards of **cl** words. Have the child classify them as action words or naming words. List the two sets of words on the chalkboard or dry-erase board. Have the child match flash cards to printed words. (Naming words [nouns]: *cloud, clam, class, clay, clothes, clock, club*. Action words [verbs]: *clean, climb, clap, claim, clang, clip*.)
- Have the child dictate sentences about a Clown Club clown using **cl** words from the lists of action and naming words. (Example: *Cleo Clown wears funny clothes*.) Have the child trace and/or copy their sentences, then illustrate them for a Clown Club storybook.

Additional Resources:
- Branley, Franklyn M. *Down Comes the Rain*. New York: HarperCollins Publishers, 1997.
- De Paola, Tomie. *The Cloud Book*. New York: Holiday House, Inc., 1975.

Published in 2002 by The Rosen Publishing Group, Inc.
29 East 21st Street, New York, NY 10010

Copyright © 2002 by The Rosen Publishing Group, Inc.

All rights reserved. No part of this book may be reproduced in any form without permission in writing from the publisher, except by a reviewer.

Book Design: Haley Wilson

Photo Credits: Cover © Allen Russell/Index Stock; p. 3 © Paul Gallaher/Index Stock; p. 5 © Mitch Diamond/Index Stock; p. 7 © Jeffry Myers/Index Stock; p. 9 © Michael Agliolo/International Stock; p. 11 © Wendy Shattil/Index Stock; p. 13 © Sandi Langley/Index Stock; p. 15 © Michele Burgess/Index Stock; p. 17 © Richard Hamilton Smith/FPG International; p. 19 © Benelux Press/Index Stock; p. 21 © VCG/FPG International.

Library of Congress Cataloging-in-Publication Data

Tanner, Susan, 1965-
   Clouds : learning the CL sound / Susan Tanner.— 1st ed.
     p. cm. — (Power phonics/phonics for the real world)
   ISBN 0-8239-5942-2 (lib. bdg. : alk. paper)
   ISBN 0-8239-8287-4 (pbk. : alk. paper)
   6-pack ISBN 0-8239-9255-1
   1. Clouds—Juvenile literature. [1. Clouds.]   I. Title.
   II. Series.
QC921.35 .T36 2002
551.57'6—dc21
                                     2001000085

Manufactured in the United States of America